Health
Grade Six

Table of Contents

Introduction. 1	Eye Traits 26
Curriculum Correlation 2	Dominant and Recessive Traits 27
Assessment. 3	Genetic Chances 28
	Variation in Traits 29
Unit 1: Notice Nutrition	Comparing Human Traits 31
The Food Guide Pyramid 5	Heredity or Environment. 32
A Balanced Diet 7	Comparing Me!. 33
To Be a Vegetarian 9	
Calories. 11	**Unit 3: Healthy Habits**
Nutrients and Calories. 12	The Cost of Smoking 34
Read a Food Label 13	The Effects of Tobacco Smoke 35
Cavities. 14	Views About Smoking. 36
Sugar and Cavities 15	The Effects of Alcohol 37
The Digestive System 17	Look at a Medicine Label 39
Digestion and Circulation 18	A Drug-Safe Home 40
Cells Obtain Food 19	Kinds of Drugs 41
Magazine Ad. 20	Reverse Crossword Puzzle 42
	Stress on Your Body 43
Unit 2: It's Ap-parent!	Identifying a Stressful Situation 44
What Do You Know About Heredity? 21	On Target for Healthy Habits. 45
The Father of Genetics 22	Good Hygiene. 47
Heredity and Genes. 23	
Chromosomes 24	**Answer Key** 48
Your Traits 25	

The school curriculum designates the topics taught to students, with the most stress placed on the core subjects of language arts and math. Health is the subject usually taught in the spare days between units in order to meet standards. However, health teaches students valuable life skills that they need to learn to develop healthy living habits. These skills will help students make responsible choices that affect their daily lives.

In this series, *Health,* information was compiled to supplement the required standards for health in each grade level. The activities were selected to complement the core subjects. So, instead of fitting health into the curriculum in one or two specific units, it can be introduced in any subject throughout the year. (The Curriculum Correlation chart on page 2 will help you determine which activities to incorporate into your core subjects.) The activities are fun and challenging, thereby avoiding the stigma of "boring" health worksheets. Many activities are open-ended so that students will have to think about, evaluate, and apply healthy habits to their daily lives. Moreover, introducing the activities into the core classes may reinforce positive habits.

Health is divided into three units. Unit 1 provides information on nutrition. It includes details about the Food Guide Pyramid, food nutrients, and ways different body systems break down and diffuse the food. Unit 2 explores heredity. Here, students learn about recessive and dominant genes, as well as determining physical appearance. Unit 3 introduces the importance of lifelong health habits, such as exercising, avoiding drugs, and identifying stressful situations. As students learn about health more often, it serves as a reminder of healthy habits. They will more likely remember the information and apply it to their daily lives—the goal of any health program.

Health
Grade Six
Curriculum Correlation

Activity	Math	Language Arts	Social Studies	Science
Unit 1: Notice Nutrition				
The Food Guide Pyramid	number sense use charts	comprehension use pictures		
A Balanced Diet	number sense use charts	comprehension use pictures		
To Be a Vegetarian	number sense	comprehension		
Calories	measurement use charts		research tools	energy
Nutrients and Calories		use pictures		
Read a Food Label	percentage number sense measurement	comprehension		
Cavities		composition		experiment
Sugar and Cavities	measurement use charts	plays		matter
The Digestive System		comprehension use pictures		systems
Digestion and Circulation		use pictures	research tools	systems
Cells Obtain Food		sequence use pictures	research tools	systems
Magazine Ad		composition		
Unit 2: It's Ap-parent!				
What Do You Know About Heredity?			family	
The Father of Genetics	use charts		family history	plants
Heredity and Genes	use charts probability	comprehension	family	
Chromosomes		use pictures	family	experiment
Your Traits	use charts	comparison	family	
Eye Traits		comprehension use pictures	family	
Dominant and Recessive Traits	use charts collect data			
Genetic Chances	use charts probability		family	
Variation in Traits	measurement collect data use graphs			
Comparing Human Traits	number sense	use pictures		
Heredity or Environment			family	environment
Comparing Me!		comparison composition	family	
Unit 3: Healthy Habits				
The Cost of Smoking	computation money time			
The Effects of Tobacco Smoke		composition	social issues	experiment
Views About Smoking	collect data use charts		social issues	
The Effects of Alcohol	decimals measurement	vocabulary comprehension	social issues	
Look at a Medicine Label	measurement			
A Drug-Safe Home	use charts		family	
Kinds of Drugs		composition	research tools social issues	chemistry
Reverse Crossword Puzzle		vocabulary	research tools social issues	chemistry
Stress on Your Body	use charts	use pictures	social issues	
Identifying a Stressful Situation	use charts	use pictures	social issues	
On Target for Healthy Habits	use charts		safety	
Good Hygiene	use charts		social issues	

www.svschoolsupply.com
© Steck-Vaughn Company

Health 6, SV 2854-1

Name _____ Date _____

Health Assessment

Darken the circle beside the answer that correctly completes each statement.

1. The Food Guide Pyramid shows the six food groups and the number of servings needed to eat a _____ diet.
   - Ⓐ vegetarian
   - Ⓑ balanced
   - Ⓒ high-fat
   - Ⓓ protein

2. The Food Guide Pyramid shows that people need to eat more foods from the _____ group.
   - Ⓐ Fats, Oils, Sweets
   - Ⓑ Milk, Yogurt, Cheese
   - Ⓒ Bread, Cereal, Rice, Pasta
   - Ⓓ Meat, Poultry, Fish

3. A _____ is the unit that measures the amount of energy in food.
   - Ⓐ Tablespoon
   - Ⓑ Decibel
   - Ⓒ Decameter
   - Ⓓ Calorie

4. Cavities are holes in the teeth caused by _____ .
   - Ⓐ milk
   - Ⓑ vinegar
   - Ⓒ bacteria
   - Ⓓ nicotine

5. Nutrients move into the bloodstream through the _____ .
   - Ⓐ small intestine
   - Ⓑ stomach
   - Ⓒ large intestine
   - Ⓓ esophagus

6. _____ is the passing on of traits from parents to children.
   - Ⓐ Digestion
   - Ⓑ Heredity
   - Ⓒ Circulation
   - Ⓓ Traitors

7. Gregory Mendel is known as the Father of _____ .
   - Ⓐ Genetics
   - Ⓑ Astrology
   - Ⓒ Children
   - Ⓓ History

Go on to the next page.

Name _____ Date _____

Health Assessment, p. 2

8. Strong genes that influence how a person looks are _____ genes.
   - Ⓐ recessive
   - Ⓑ dominant
   - Ⓒ protective
   - Ⓓ chromosome

9. Heredity and the _____ help to shape a person's abilities.
   - Ⓐ environment
   - Ⓑ muscular system
   - Ⓒ bones
   - Ⓓ genes

10. The _____ are the part of the body most affected by cigarette smoke.
    - Ⓐ nerves
    - Ⓑ muscles
    - Ⓒ eyes
    - Ⓓ lungs

11. Blood alcohol content (BAC) means the amount of alcohol per 100 units of _____.
    - Ⓐ oxygen
    - Ⓑ nerves
    - Ⓒ blood
    - Ⓓ saliva

12. _____ is the disease in which a person is unable to control his or her drinking.
    - Ⓐ Cirrhosis
    - Ⓑ Cancer
    - Ⓒ Alcoholism
    - Ⓓ Emphysema

13. Drugs that speed up the functions of the body are called _____.
    - Ⓐ stimulants
    - Ⓑ amphetamines
    - Ⓒ uppers
    - Ⓓ all of the above

14. All of your health habits are called a _____.
    - Ⓐ movement
    - Ⓑ meeting
    - Ⓒ system
    - Ⓓ lifestyle

Name _____ Date _____

The Food Guide Pyramid

The Food Guide Pyramid shows you how to keep your diet balanced. It shows the six food groups, the kinds of foods in each, and the number of servings of each food group you should eat each day to keep a balanced diet. Study the Food Guide Pyramid below.

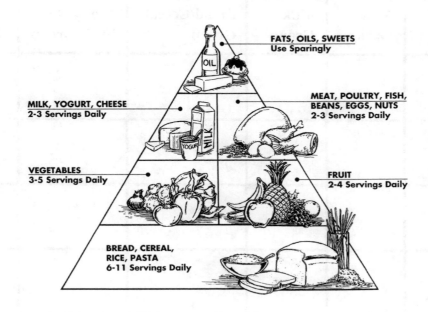

Answer these questions.

1. What are some of your favorite foods? _____
   _____

2. From which group do most of your favorite foods come? _____

3. Do you think that you have a well-balanced diet? _____

4. Do you eat too many foods from the Fats, Oils, and Sweets group? _____

5. How do you think you can improve your eating habits? _____
   _____

Go on to the next page.

Name _____ Date _____

The Food Guide Pyramid, p. 2

You have heard the saying, "You are what you eat." Do you agree with that saying? In the chart below, list the foods you ate yesterday. Write the name of each food under its food group and across from the nutrient it contains. You may need to write a food in more than one place.

Nutrient	Meat Group	Milk Group	Bread/Cereal Group	Vegetable/ Fruit Group	Fats/Oils/ Sweets Group
Carbohydrate					
Protein					
Fat					
Vitamins					
Minerals					

When you have completed the chart, study it to find out if yesterday's diet was a balanced diet.

Answer these questions.

1. From which nutrient group did most of your food come?

   _____

2. Did you eat foods from each food group? _____

3. How many servings did you have from each group? _____

   _____

4. Was your diet balanced? _____

Name _____ Date _____

A Balanced Diet

....

Scientists have studied the amounts of foods that are necessary for people to eat in order to get enough nutrients, vitamins, and minerals. The Food Guide Pyramid below shows how much of each type of food people should eat each day to stay healthy.

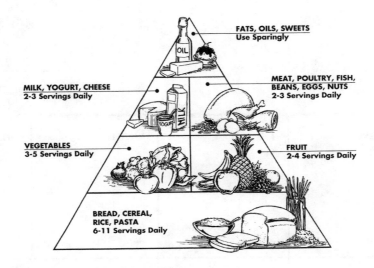

Using the information provided in the food pyramid, make a menu for one day. Be sure that your menu includes the number of servings that are recommended in the pyramid.

BREAKFAST	LUNCH	DINNER	SNACKS
Food Type	Food Type	Food Type	Food Type

Total your servings for each group. Does your menu follow the pyramid recommendations?

Bread/Cereal Group: _____ Vegetable Group: _____

Fruit Group: _____ Milk Group: _____

Meat Group: _____ Fats Group: _____

Go on to the next page.

Name _____ Date _____

A Balanced Diet, p. 2

Keep track of what you eat for three days. Then, compare your diet with the diet that is recommended on the Food Guide Pyramid.

Day 1	Breakfast	Lunch	Dinner	Snacks
Day 2	Breakfast	Lunch	Dinner	Snacks
Day 3	Breakfast	Lunch	Dinner	Snacks

Answer these questions.

1. Which food do you need to eat more of? _____

2. Which food do you need to eat less of? _____

3. Do you have a healthy diet? _____

To Be a Vegetarian

You've probably heard about vegetarians, but you may not know exactly what being a vegetarian means. Someone who is a vegetarian follows a vegetarian diet and eats no meat of any kind. That means eating no flesh of any animal, including chicken and fish.

People usually follow a vegetarian diet for one of two reasons—either for moral reasons or for health concerns. Many people feel that it's not right to eat other animals. They may also feel that eating meat is a wasteful practice when so many people in the world do not have enough to eat. This reasoning is based on the fact that cattle are fed grain that could be used to feed people. As you know from reading about food webs, energy is lost as it goes from one living thing to the next. In other words, a cow takes in much more energy than a human or another animal could gain from eating its flesh.

Many people follow a vegetarian diet because of health concerns. A vegetarian diet is one way of lowering the amount of fat a person eats. The traditional American meat-centered diet is very high in fat. And high-fat diets have been shown to be a factor in many diseases, including heart disease and cancer.

Some vegetarians avoid not only meat but also animal products, such as eggs and milk. This diet is harder to follow because many foods, from breads to salads, may contain ingredients made from milk or eggs. Other vegetarians believe it is okay to eat animal products as long as the animal is not harmed. These people get protein from foods such as yogurt, cottage cheese, and other milk products. Some vegetarians also eat products that contain eggs.

Vegetarians who eat egg and milk products can easily get enough protein, an important part of a healthful diet. Those who do not eat any eggs or milk must be more careful to make sure they get enough protein in their diets. Many foods, such as beans, nuts, and grains, have high levels of protein. But a variety of these foods must be eaten together to provide the proteins the body needs to stay healthy. Eating beans and rice together provides this kind of protein.

Go on to the next page.

Name _____ Date _____

To Be a Vegetarian, p. 2

In the past, it was thought that people needed to eat large amounts of protein. Now, research has shown that people need less protein and more grains, fruits, and vegetables. The new Food Guide Pyramid shows the suggested amount of different foods. The large base of the pyramid shows those foods needed in the greatest amounts. As you go up the pyramid, you find foods that you need in smaller and smaller amounts.

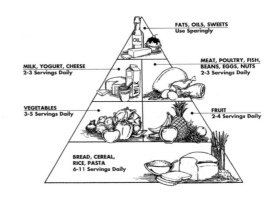

Answer these questions.

1. What is a vegetarian?

   _____

   _____

2. Why might it be more healthful to be a vegetarian than to eat a traditional American diet?

   _____

   _____

3. According to the Food Guide Pyramid, how many servings of bread or cereal should you eat each day? How many servings of vegetables? Of fruits?

   _____

   _____

4. Suggest a menu for a vegetarian meal. If you like, it may contain milk and egg products.

   _____

   _____

   _____

   _____

Name _____ Date _____

Calories

A *Calorie* is the unit that measures the amount of energy in food. Fruits and vegetables have smaller amounts of Calories, while desserts have more.

Knowledge about the food you eat is important if you wish to have a balanced diet. You will need diet and health books to obtain this information.

A. Choose six of your favorite foods, and list them in the chart below. Draw a picture of the food, and write down the food group it belongs to.

B. Using the diet and health books, find out the number of Calories in one serving of each food item. Are you consuming more Calories than you need?

Food Item	Picture	Food Group	Number of Calories

Name _____   Date _____

Nutrients and Calories

Your body needs nutrients in order to grow and stay healthy. A balanced diet provides all the nutrients you need.

A. Match the nutrients listed in the left-hand column with the uses listed in the right-hand column.

1. Carbohydrates
2. Fats
3. Protein
4. Vitamins
5. Minerals
6. Water

a. Growth and repair
b. Small amounts for growth and activity
c. To form parts of the body
d. Energy
e. Makes up 75 to 80% of the body
f. Energy storage

B. Name a food source for each nutrient listed above.

_____
_____
_____

C. Look at the pictures below. Which person would use more Calories? Explain.

_____
_____
_____

Name _____ Date _____

Read a Food Label

The skill of observation is important when deciding what foods to eat. Read the cereal box shown in the margin, and answer the questions below.

1. How many Calories are in one serving of this cereal without milk? With milk?

   _____

   _____

2. Does this cereal contain more carbohydrates or more fats?

   _____

3. Does this cereal contain more starch or more sugar?

   _____

4. What does RDA mean?

   _____

   _____

5. How much of the U.S. RDA of vitamin A is in $\frac{1}{4}$ cup of cereal?

   _____

6. What minerals are in this cereal?

   _____

   _____

7. Which do you think are the five main ingredients in this cereal?

   _____

   _____

8. According to the cereal box, how could you increase the amount of protein in this cereal?

   _____

   _____

Each serving contains 4 g dietary fiber, including 1 g (2.6% by weight) non-nutritive crude fiber.

NUTRITION INFORMATION PER SERVING

Serving Size: 1/4 cup Raisin Bran (1 ounce bran flakes with 1/3 ounce raisins) alone, and in combination with 1/2 cup Vitamin D fortified whole milk.
Servings Per Container: 15

	Raisin Bran	
	1 oz. Cereal & 1/3 oz. raisins	with 1/2 cup whole milk
Calories	120	190
Protein	3 g	7 g
Carbohydrate	29 g	35 g
Fat	1 g	5 g

PERCENTAGE OF U.S. RECOMMENDED DAILY ALLOWANCE (U.S. RDA)

	Raisin Bran	
	1 oz. Cereal & 1/3 oz. raisins	with 1/2 cup whole milk
Protein	4	15
Vitamin A	25	30
Vitamin C	*	2
Thiamin	25	30
Riboflavin	25	35
Niacin	25	25
Calcium	*	15
Iron	25	25
Vitamin D	10	25
Vitamin B4	25	25
Folic Acid	25	25
Vitamin B12	25	30
Phosphorus	15	25
Magnesium	15	20
Zinc	25	30
Copper	6	6

*Contains less than 2% of the U.S. RDA of these nutrients.

INGREDIENTS: Wheat Bran with other parts of wheat; Raisins; Sugar; Salt; Malt Flavoring; Partially Hydrogenated Vegetable Oil (One or More of: Coconut, Soybean, and Palm); Invert Syrup; Vitamin A Palmitate; Reduced Iron; Zinc Oxide; Niacinamide; Pyridoxine Hydrochloride (B6); Thiamin Hydrochloride (B1); Riboflavin (B2); Folic Acid; Vitamin B12; and Vitamin D2.

CARBOHYDRATE INFORMATION

	Raisin Bran	
	1 oz. Cereal & 1/3 oz. raisins	with 1/2 cup whole milk
Starch and Related Carbohydrates	12 g	12 g
Sucrose and Other Sugars	13 g	19 g
Dietary Fiber	4 g	4 g
Total Carbohydrates	29 g	35 g

Values by Formulation and Analysis.

Name _____ Date _____

Cavities

Cavities are holes in the teeth that are caused by bacteria. Dentists tell us to brush, floss, and rinse our teeth every day to get rid of the bacteria. How can taking care of our teeth prevent cavities? This activity will give you the answers.

Materials:
eggshells
a cup of vinegar
a cup of water

Do this:
A. Place two pieces of eggshell in a cup of water.
B. Place two pieces of eggshell in a cup of vinegar.
C. Set the cups aside for three days. Then, remove the eggshells from the cups. Compare them.

How do the eggshells that were in the water feel? _____

How do the eggshells that were in the vinegar feel? _____

Eggshells and teeth both contain calcium. The calcium makes them hard. Vinegar, an acid, ate away at the calcium in the eggshells. The bacteria in your mouth act on the sugar in foods, and acids form. The acids work on your teeth in the same way vinegar worked on the eggshells.

There is a way to stop acid from eating away at your teeth. All you have to do is brush, floss, and rinse after you eat. This gets rid of the bits of food left in your mouth. Then, the bacteria cannot produce acids that can break down the enamel covering of your teeth.

Write a television jingle that will convince people to brush their teeth.

_____
_____
_____
_____
_____
_____

Name _____ Date _____

Sugar and Cavities

Conversation at a Dentist's Office

Rita: Guess what! No cavities!

Kelly: That's not fair! I have three cavities! And I brush my teeth every day. I don't get it.

Dr. Young: Brushing your teeth isn't the only way to prevent cavities. If you eat a lot of sugar, you're inviting cavities. Sugar is rapidly broken down in your mouth by bacteria. The bacteria produce acids that begin the decaying process.

Kelly: But I don't eat that much sugar!

Dr. Young: You may eat more sugar than you think. Remember, sugar is contained in a lot of foods. Take a look at some food labels. Look for not only the word *sugar* but also for other names for sugar, such as *honey, molasses, sucrose, corn syrup, dextrose, fructose,* and *corn sweeteners.*

Why don't you keep a food diary for a day? Write down everything you eat, not only during meals but for snacks as well. Now, listen carefully to this: The problem is not just the *amount* of sugar eaten daily. It is also *how often* you eat sugar and *how long* it stays in your mouth. *The greatest damage is done within the first 20 minutes after eating sweet foods.* The more times you eat sugar, the more times acid forms on your teeth. So at the end of the day, review your food diary. Circle all meals and snacks you ate that contained sugar. Multiply the number of times you ate sugar by 20. That is how long acid was working on your teeth.

Go on to the next page.

Name _____   Date _____

Sugar and Cavities, p. 2

Here are Kelly's and Rita's food diaries. Calculate how many minutes acid was attacking their teeth. Then, try keeping your own food diary.

DIET RECORDS		
	Kelly	Rita
Breakfast	cereal with milk and sugar toast with jam	oatmeal with milk 100% orange juice toast with butter
Snack	glazed donut	apple
Lunch	soda peanut butter and jelly sandwich potato chips chocolate chip cookie	milk fresh salad cheese sandwich fresh peach
Snack	candy bar	peanuts
Snack	gum	
Dinner	chocolate milk glazed ham peas potato ice cream	milk chicken potato beans fresh salad ice cream
Snack	milk cake with frosting	popcorn

Minutes: _____   _____

• Another important note: Gum disease is actually the number-one cause of tooth loss. Brushing <u>and</u> flossing are vital to healthy gums.

The Digestive System

Digestion is the process in which food is broken down into nutrients so that our body can get energy. As food enters the mouth, the **teeth** grind it up. Saliva, the fluid in the mouth, has chemicals that begin to break down the starch in foods. After we swallow, the food moves down the **esophagus** to the **stomach.** More chemicals mix with the food to break down the food until it is a liquid.

The food then moves to the **small intestine.** Such organs as the **liver, gallbladder,** and **pancreas** add other chemicals to the small intestine to break down other nutrients, like fat. Many important nutrients now leave the small intestine and seep into the bloodstream. Other food that cannot be totally broken down moves to the **large intestine.** It stays there until the waste is removed.

Look at the words in dark print above. Find and label each part on the picture below.

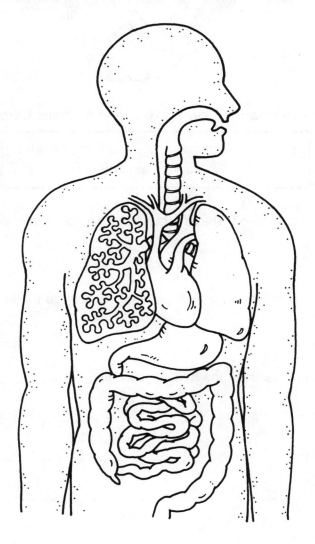

Name _____ Date _____

Digestion and Circulation

A. In the diagram below, label the four chambers of the heart. Then, trace the flow of blood through it. You may refer to a science book for help.

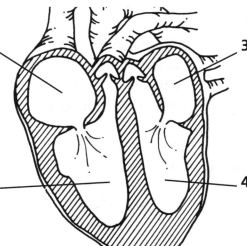

1. _____

2. _____

3. _____

4. _____

B. Use the words from the box to complete the sentences below.

> oxygen   diffusion   veins   cell membranes
> villi   capillaries   arteries

5. The blood vessels that carry blood from the heart to all parts of the body are the

   _____.

6. The blood vessels that carry blood from all parts of the body to the heart are the

   _____.

7. The blood travels from the heart to the lungs to pick up _____.

8. Dissolved nutrients pass through the _____ lining the small intestine and

   enter the bloodstream through the tiny _____. The nutrients enter

   the cells through their _____ by a process called

   _____.

Name _____ Date _____

Cells Obtain Food

The pictures show the places in the path digested food travels from your small intestine to your body cells. The pictures are out of order. Use a reference book to help you complete this exercise.

1. Number the pictures in the correct order from 1 to 5.
2. Draw an arrow that shows in which direction the blood and food are moving in each picture.
3. Place a star beside any picture that shows diffusion.

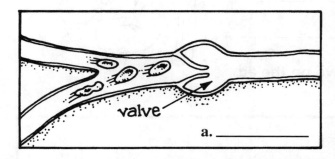

a. _____

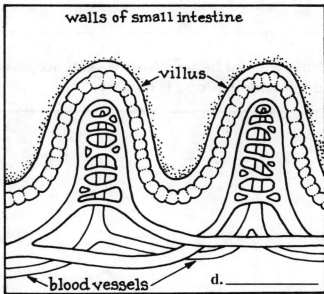

d. _____

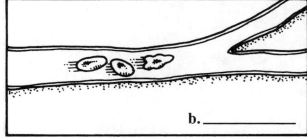

b. _____

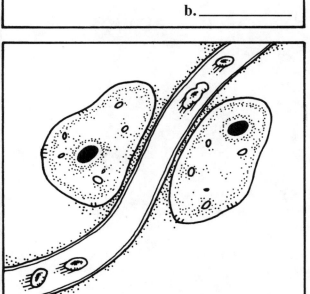

c. _____

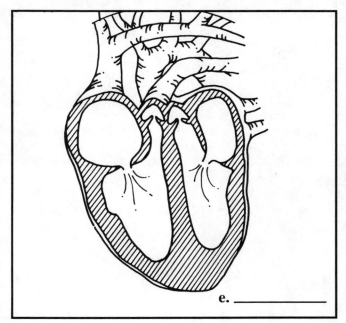

e. _____

Name _____ Date _____

Magazine Ad

Suppose you work for an advertising company. You are to design an ad for a new, healthy food.

What is the name of the food? _____

Why is it healthy? _____

_____

_____

Draw your ad below. Write a sentence or two on the ad to tell why the food is healthy.

Name _____ Date _____

What Do You Know About Heredity?

••

Heredity is the passing on of traits from parents to children. Many people believe things about heredity without really knowing whether they are true. Find out how much you know.

Read each statement. If the statement is true, write *T* in the blank. If the statement is false, write *F* in the blank.

1. _____ Each parent gives one half of a child's genetic makeup.

2. _____ Before birth, inherited traits may be changed by the stars, the Moon, or the planets.

3. _____ A child may have traits that the parent doesn't show.

4. _____ Many of a person's inherited traits cannot be seen.

5. _____ If a person loses an arm in an accident, his or her children may be born with only one arm.

6. _____ If a woman wants her child to be an artist, she should take up painting before the child is born.

7. _____ Some inherited traits are affected by a person's blood.

8. _____ A child may have a trait that is a blend of the parent's traits.

9. _____ A child will be more like a parent who is strong-willed than one who is not.

10. _____ Parents who are tall will always have children who are tall.

Name _____ Date _____

The Father of Genetics

Gregor Mendel is known as the Father of Genetics. Because of his work, scientists know that there are two heredity factors for every trait, one of which comes from each parent. They also know that of these genes, some are dominant and some are recessive.

Dominant Traits	Recessive Traits
tall	short
round seeds	wrinkled seeds
yellow seeds	green seeds
flowers along entire stem	flowers at top of stem

The table above shows dominant and recessive traits in pea plants. Use this table to predict traits that would appear in each of the pea-plant crosses described below. Write your answers in the "Offspring" boxes.

1. **Plant 1** × **Plant 2** = **Offspring**
   tall, wrinkled seeds, yellow seeds, flowers along entire stem
   tall, round seeds, yellow seeds, flowers at top of stem

2. **Plant 1** × **Plant 2** = **Offspring**
   short, yellow seeds, round seeds, flowers along entire stem
   tall, green seeds, wrinkled seeds, flowers along entire stem

3. **Plant 1** × **Plant 2** = **Offspring**
   tall, round seeds, yellow seeds, flowers along entire stem
   short, wrinkled seeds, green seeds, flowers at top of stem

Heredity and Genes

All living things must reproduce, or make more living things like themselves. If a species did not reproduce, all living things of its kind would die out. The reproductive system of humans allows people to make more humans, or to have children.

When people have children, they pass on certain traits and characteristics. This is called *heredity*. Heredity affects the way you look and the way you act. You may have noticed that when adults look at a new baby, they often talk about which parent the baby looks like. This is because the baby has inherited its looks from its parents. As a child grows, there may be times when the child looks more like the mother, and times when the child looks more like the father. There will be certain things the child does that will remind people of the mother or the father, or even of some other relative. All these things are inherited. Other things, such as likes and dislikes and personal fitness, are not inherited. These are the result of the person's lifestyle and environment.

When you study cells, you learn that the nucleus of a cell contains chromosomes. On the chromosomes are genes. Genes determine how offspring will look and act. Each child receives genes from both parents, but some genes are stronger than others are. These genes are called *dominant*. The weaker genes are called *recessive*.

Here is an example. The gene for brown hair is a dominant gene. We say that brown hair is a dominant trait. Blond hair is a recessive trait. If both parents have brown hair, their children will probably have brown hair. If one parent has brown hair and the other parent has blond hair, the children will still most likely have brown hair, but it is possible for a child to have blond hair. If both parents are blond, then the child will probably be blond.

The combinations can be seen in a chart like this. A brown-haired mother may carry a blond-haired gene, because the brown-haired gene will dominate. She may pass on genes like this: Bb.

A blond-haired father cannot have a brown-haired gene. He must pass on genes like this: bb.

To see what combinations of genes the children can receive, we can make a chart.

	B	b
b	Bb	bb
b	Bb	bb

In this family, it is possible that half of the children could have blond hair, or there is a 50% chance that a child could have blond hair.

Suppose a brown-haired father has genes that are BB. A blond-haired mother has genes that are bb. Work the chart below to find the combinations of genes the children can receive. Then answer the questions.

1. What is the chance that a child will have brown hair? _____

2. What is the chance that a child will have blond hair? _____

	B	b
b		
b		

Name _____ Date _____

Chromosomes

Chromosomes are parts of cells that have the information telling how a person will grow and develop. Think of the two strips shown below as chromosome models. The upper one is female, and the lower one is male. You can use a model to figure out what happens when the chromosomes divide to make eggs and sperm.

Materials:

white paper
crayons or markers

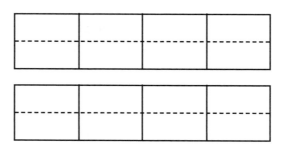

Do this:

A. Make copies of the two strips. Each rectangle on the chromosome represents one gene. Color the female chromosome red and the male chromosome blue. Cut them out.

   1. How many genes does a chromosome have for the same trait? _____

   2. How many traits are there genes for on each chromosome model? _____.

B. Cut each chromosome along the dotted line. This cut represents the division that forms eggs and sperm.

   3. How many genes does each divided chromosome have for each trait? _____

   4. How many traits are there genes for on each divided chromosome? _____

C. Match a divided red chromosome with a divided blue one. Do the gene pairs match those of either parent? _____

D. Try different combinations of red and blue divided genes. Do any combinations have the same set of genes? _____

www.svschoolsupply.com
© Steck-Vaughn Company

24

Unit 2: It's Ap-parent!
Health 6, SV 2854-1

Name _____ Date _____

Your Traits

··

You can find out about some traits you have inherited. You can also find out something about what genes you have. If you show a dominant trait, you have at least one dominant gene. If you show a recessive trait, you have two recessive genes.

A. Fill in the chart with the traits that you have. You may need a mirror to help you.

Trait		Your Trait	How Inherited
Hair — curly / straight			dominant / recessive
Earlobes — detached / attached			dominant / recessive
Tongue — roller / nonroller			dominant / recessive
Cheek — dimples / smooth			dominant / recessive
Chin — cleft / smooth			dominant / recessive

B. Determine as much as you can about your genes. Circle *dominant* if you have inherited a dominant trait. Circle *recessive* if you have inherited a recessive trait.

Name _____ Date _____

Eye Traits

You are learning about dominant and recessive traits. Dominant traits mask recessive traits. In eye color, brown is dominant over green or gray; green or gray is dominant over blue. In this activity, you will observe the eye color of members of your family and make inferences about your genes for eye color.

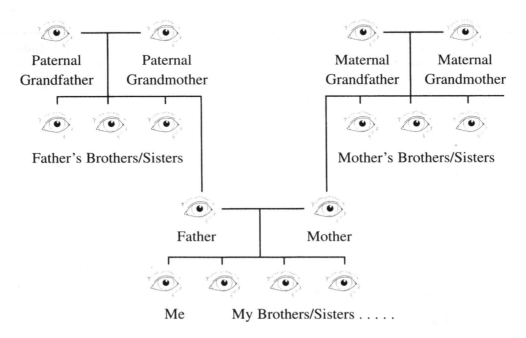

Answer these questions.

1. If all of your grandparents had two genes for the same eye color, what would have been the eye color of your father (and his brothers and sisters)?

   _____

2. Your mother (and her brothers and sisters)?

   _____

3. Were there any differences in eye color for your family? What are the possible pairs of genes for your:

   paternal grandfather? _____ paternal grandmother? _____

   maternal grandfather? _____ maternal grandmother? _____

   father? _____ mother? _____

4. What are possible gene pairs that you might have for your eye color?

Name _____ Date _____

Dominant and Recessive Traits

List how many of your classmates appear to have dominant or recessive traits for curly hair or straight hair, dimples or no dimples, and light eyes or dark eyes. Use your data to complete this chart. (Don't forget to include yourself in the count.)

Dominant Traits	Number	Recessive Traits	Number
Curly Hair		Straight Hair	
Dimples		No Dimples	
Light Eyes		Dark Eyes	

Answer these questions.

1. Which trait is the most dominant? _____

2. Which trait is the most recessive? _____

3. Do more students in your class have dominant or recessive traits?
   _____

4. Why do you think this is the case?
   _____
   _____

Name _____ Date _____

Genetic Chances

For a certain trait, you get one gene from each parent. Imagine, for example, that your mother has one gene for detached earlobes and one gene for attached earlobes. You will get one or the other of these genes. What are the chances that you will get one instead of the other? To find out, you will need a penny or other coin.

1. Flip a penny. Let "heads" stand for the gene for detached earlobes. Let "tails" stand for the gene for attached earlobes. Record the results in a chart like the one shown.

Toss	Detached Earlobes (Heads)	Attached Earlobes (Tails)
1		
2		
3		
.		
.		
.		
50		
Total		

2. Flip your coin 49 more times. Each time, record your results.

   How many "heads" came up? _____

   _____

   How many "tails" came up? _____

   _____

3. Find out the chances of getting one or the other gene. Divide each total by 50. (For example: 25 detached earlobes in 50 tries is $\frac{25}{50}$ or $\frac{1}{2}$.)
   What are the chances for detached earlobes?

   _____

   What are the chances for attached earlobes?

   _____

Name _____ Date _____

Variation in Traits

People are alike in certain ways. For example, they have eyes, noses, and ears. They are different in many ways, too. The students in your class may differ in height, weight, hair color, eye color, and facial appearance. You can examine one of these traits, height, to see just how much variation there is.

Materials:
meter stick
graph paper

Do this:
1. Have one of your classmates measure your height. Record your height in a class chart the teacher compiles.
2. Look at the class height chart. Then, in the chart below, record the number of boys and the number of girls in each height range. Also record the total number of students in each height range.

Height (in cm)	Number of Girls	Number of Boys	Total Number of Students
below 134			
135–144			
145–154			
155–164			
above 165			

3. Make three graphs of the information in your chart—one for girls, one for boys, and one for the total. Set up each graph as shown on the next page.

Go on to the next page.

Name _____ Date _____

Variation in Traits, p. 2

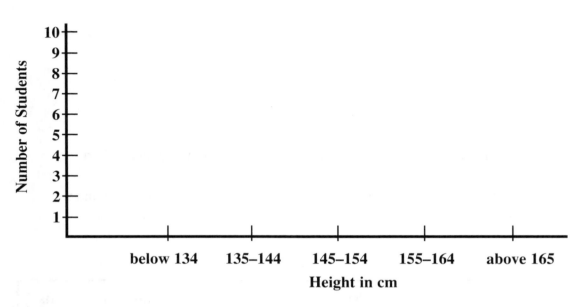

Height Graph for _____

Study the graphs. Then, answer these questions.

1. What shape do the graphs take? Explain. _____
   _____
   _____

2. How do the heights of the boys and girls compare? _____
   _____

   Why is this probably true? _____
   _____

Name _____ Date _____

Comparing Human Traits

A. Use the chart to find your combination of the five traits.

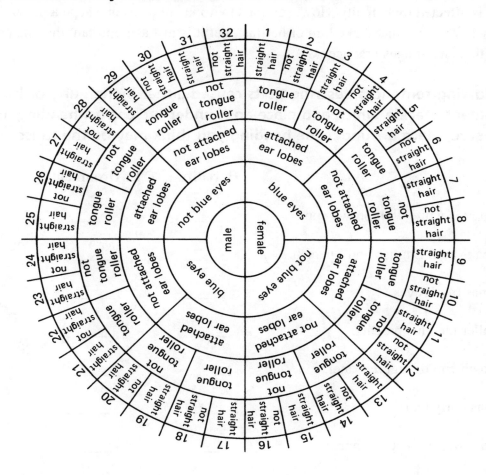

B. Locate *male* or *female* on the chart. Move out one ring to eye color. Continue until you reach the number on the outside edge.

1. What is the number for your combination of five traits?

   _____

2. How many other students have the same trait number?

   _____

3. How many have different trait numbers?

   _____

www.svschoolsupply.com 31 Unit 2: It's Ap-parent!
© Steck-Vaughn Company Health 6, SV 2854-1

Name _____ Date _____

Heredity or Environment

Heredity affects how a person looks and grow. Sometimes a person's ability to do activities, like running, is affected by heredity. However, the environment also can shape a person's abilities. A runner may have genes that make him or her fast, but it is just as important that the person has an environment that encourages practice.

A. The following sentences describe traits caused either by heredity or by environment. Decide which and place an *X* under the correct heading. If you are not sure, place a ? under the heading you think may be the cause.

	Heredity	Environment	Both
1. This puppy is white with brown spots.	_____	_____	_____
2. Tina speaks both English and Spanish.	_____	_____	_____
3. Miguel has black eyes.	_____	_____	_____
4. Jim is taller than his parents.	_____	_____	_____
5. Beth is a champion figure skater.	_____	_____	_____
6. Gloria has curly hair.	_____	_____	_____
7. Secretariat was a great racehorse.	_____	_____	_____
8. Nolan Ryan threw the fastest pitch.	_____	_____	_____
9. Lucy is the fastest runner in her class.	_____	_____	_____
10. A well-fed mouse looks fat and healthy.	_____	_____	_____

B. Compare your answers with those of your classmates. Discuss the reasons for your choices. Would you change any of your answers?

Name _____ Date _____

Comparing Me!

Think about how you and one of your parents are alike and different. Then, complete the Venn diagram to show the characteristics.

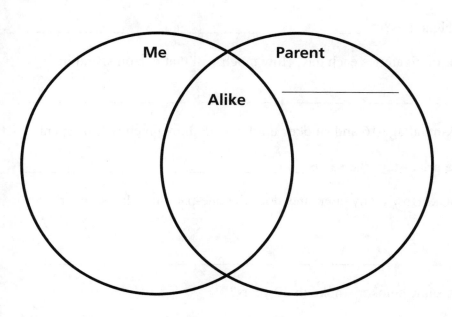

Write a compare and contrast paragraph to show the information on the Venn diagram.

_____
_____
_____
_____
_____
_____
_____
_____
_____

Name _____ Date _____

The Cost of Smoking

Smoking tobacco is not only dangerous to your health, but it is also expensive. Answer the questions and find out why.

1. What is the cost of a pack of cigarettes? _____

2. Say a person smokes one pack of cigarettes each day. How much will that person spend in one year? _____

3. If the same person starts smoking at age 16 and smokes until age 55, how much will be spent on cigarettes? Assume that the price stays the same. _____

4. There are 20 cigarettes in a pack. How many cigarettes does the one-pack-per-day smoker smoke in one year?

   _____

5. How many cigarettes does the same smoker smoke in 40 years?

   _____

6. If one cigarette shortens a smoker's life by 5.5 minutes, by how many minutes has the smoker shortened his or her life?

   _____

7. By how many years has the smoker's life been shortened?

   _____

8. Can you think of other reasons that smoking costs money?

   _____

Unit 3: Healthy Habits
Health 6, SV 2854-1

Name _____ Date _____

The Effects of Tobacco Smoke

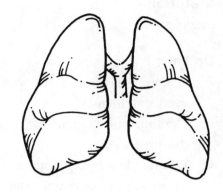

Materials:
a piece of white cotton cloth
an adult who smokes cigarettes

Do this:
A. Ask the adult to inhale the smoke from a cigarette. Then, have the person stretch the white cloth across his or her mouth and exhale the smoke through the cloth.
B. Look at the cloth.

Answer these questions.

1. How has the cloth changed? _____
_____

2. After seeing what smoke does to the cloth, what do you think it does to the inside of the smoker's lungs?
_____
_____
_____

3. Is the comparison between the cloth and human beings a good one? Explain. _____
_____
_____

4. Write a persuasive paragraph that could encourage a smoker to quit.
_____
_____
_____
_____
_____

Name _____ Date _____

Views About Smoking

Material:

graph paper

Do this:

A. Ask at least three adults who smoke and three adults who do not if they agree or disagree with each statement.

B. Combine all the data. Record the number of smokers and nonsmokers who agreed and disagreed.

C. Use the data to construct a bar graph. Put statements along a vertical axis and the number of people agreeing with statements on a horizontal axis. Label one key SMOKERS and one key NONSMOKERS.

	Smokers		Nonsmokers	
Statements	Agree	Disagree	Agree	Disagree
1. Smoking is a harmful habit.				
2. It is OK to smoke only five cigarettes a day.				
3. If you are healthy, smoking will not harm you.				
4. Tobacco smoke harms the people around smokers.				

Answer these questions.

1. How many agreed and disagreed?

_____

_____

2. Read the statements. How much difference was there between the responses of smokers and nonsmokers?

_____

_____

Name _____ Date _____

The Effects of Alcohol

When a person drinks alcohol, the alcohol is absorbed into the system very quickly and travels to the brain. If drinks are taken too quickly, the blood alcohol content (BAC) is raised. *BAC* means the amount of alcohol per 100 units of blood. The chart below shows what happens to the BAC as a person drinks. When this content is very high, it means the person is "drunk." A person should not drive a car or do anything that might cause danger to oneself or to others. Sometimes when a person drinks too much, the next day he or she has a hangover.

Alcoholism is a disease. Alcoholics are not able to control their drinking. They are often drunk, which hurts themselves, their families, and their friends. They may also hurt their health. Sometimes they end up with a condition called cirrhosis, a scarring of the liver, which can lead to death.

There are times when even a little alcohol can be harmful to a person. For example, if a pregnant woman drinks, it can cause fetal alcohol syndrome (FAS) in her baby. FAS can cause a variety of birth defects in a baby.

It is important to remember that drinking does not make a person a grown-up. It is all right to choose not to drink. It is important not to be pressured into drinking if you don't want to. Remember, the choice is yours!

BAC for a Person Weighing 120 Pounds		
Number of Drinks in One Hour	BAC	Effects
1 2	.04 .08	Alcohol affects thinking and emotions. Good judgment and self-control decrease.
3 4 5	.11 .15 .19	It becomes difficult to think clearly. Even simple movements become difficult.
6 7 8	.23 .26 .30	Speech becomes slurred. Hearing and vision are impaired.
9 10	.34 .38	The person can no longer stand. There may be vomiting and unconsciousness.

Go on to the next page.

The Effects of Alcohol, p. 2

DOWN

1. _____ is a disease in which the person is unable to control his or her drinking.
5. Scarring of the liver, often caused by drinking, is called _____.
6. A person with a high BAC is said to be _____.

ACROSS

2. Women who drink too much during pregnancy may give birth to babies with _____ _____ syndrome.
3. The effect of alcohol on the brain is related to the body's ability to _____ it.
4. The initials that define the amount of alcohol per 100 units of blood are _____.
7. Drinking too much may cause a _____, with a headache or stomachache.

Name _____ Date _____

Look at a Medicine Label

....

Gone are the days when a medicine man pulled his wagon into town to sell useless or harmful liquids as medicine. Today, the Food and Drug Administration (FDA) sets rules for selling medicines. The FDA asks makers of medicines to include the following information about their drugs:

1. Name and description of the medicine
2. Actions or Indications—Statements about illnesses for which a drug can be used
3. Cautions—Statements about when the drug should not be taken and possible harmful side effects
4. Warnings—Statements about possible dangers from taking it
5. Dosage and Administration—Statements telling how much (dosage) to take and when to take it (administration)

On the label below, place a number (1 through 5 from above) next to the information that goes with the statements.

STOMACH MEDICINE
Shake well before using.

For indigestion
For nausea
For diarrhea
For abdominal cramps
Active Ingredient: bismuth subsalicylate
Contains no sugar

DIRECTIONS:
Adults-2 tablespoons
Children: according to age
    10 to 14 years-4 teaspoons
    6 to 10 years-2 teaspoons
    3 to 6 years-1 teaspoon
Repeat dosage every 1 hour until 8 doses are taken.

CAUTION: IF TAKEN WITH ASPIRIN AND RINGING IN THE EARS OCCURS, DISCONTINUE USE. IF YOU HAVE A FEVER FOR MORE THAN 2 DAYS, CONSULT A DOCTOR.

WARNING: KEEP OUT OF THE REACH OF SMALL CHILDREN.

Name _____ Date _____

A Drug-Safe Home

Drugs can be very helpful if used properly. Therefore, it is important to know the drugs we keep at home, where they are stored, and how they are used. Complete the survey below to find out how drug-safe your home is.

SAFETY TIP: IT IS IMPORTANT THAT THIS ACTIVITY BE SUPERVISED BY AN ADULT.

Type of Drug	Use	Where Stored

Answer these questions.

1. Are these drugs kept in a safe place? _____

2. How many of the drugs in your survey are prescription drugs? _____

   How many are nonprescription? _____

Name _____ Date _____

Kinds of Drugs

Some drugs are called "uppers" because they speed up the functions of the body. For example, uppers can cause your heart to beat faster. Uppers are also called stimulants. Some drugs are called "downers." These drugs slow down the functions of the body. Downers are also called depressants. Below is a list of uppers and a list of downers. Select one from each list, and write a report. The following questions should be answered in your report.

1. What is the drug made from?
2. What are some of the effects of the drug on the body?
3. Is use of the drug legal or illegal?
4. Are there benefits in using the drug?
5. Are there dangers in using the drug?
6. How many people are reported to be using the drug?

Uppers or Stimulants	Downers or Depressant Drugs
benzedrine	heroin
dexadrine	morphine
cocaine	codeine
caffeine	barbiturates

Notes for Report

_____
_____
_____
_____
_____
_____
_____
_____
_____
_____

Reverse Crossword Puzzle

The crossword puzzle is backward. That is, the answers are given. You have to write the clues to the answers on the lines below.

Down

1. _____
2. _____
4. _____
5. _____

Across

3. _____
6. _____
7. _____
8. _____
9. _____

Name _____ Date _____

Stress on Your Body

Did your heart ever start pounding rapidly from anger? This is your body's response to stress. Stress is a normal part of life. Stress can be caused by many different things. An argument with a friend, a test in school, or even loud noises can be stressful.

Sometimes severe stress can make you ill. If you are constantly under stress, your blood pressure may rise or your digestion may be affected. This picture shows some illnesses caused by severe stress.

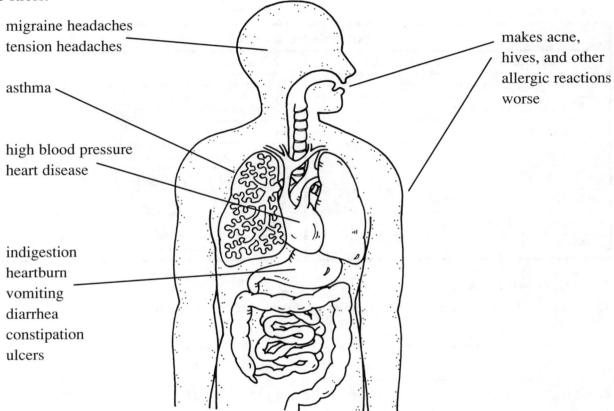

- migraine headaches
- tension headaches
- asthma
- high blood pressure
- heart disease
- indigestion
- heartburn
- vomiting
- diarrhea
- constipation
- ulcers
- makes acne, hives, and other allergic reactions worse

What can you do about too much stress? It is important to develop habits to keep your life from becoming overly stressful. Take this test. Can you answer *Yes* to all of the questions?

1. I participate in group activities at school or elsewhere.
2. I have close friends and relatives whom I can call on in times of trouble.
3. I sleep well at least 7 to 8 hours each night.
4. I am generally healthy and happy.
5. I enjoy hobbies or other relaxing activities every day.

Yes	No

Name _____ Date _____

Identifying a Stressful Situation

The chart below lists 18 possible sources of stress. As you read each source, think about what it means to you. Then, number a piece of paper from 1 to 18. Now, list the sources of stress in order from those you find to be most stressful to those you find to be least stressful. (The most stressful will be number 1.)

Possible Sources of Stress	
Not having done your homework	Going on a trip
Having an argument at home	Imagining what others think of you
Meeting new people	Being in a sports competition
Taking a test	Giving a speech
Getting sick	Moving into a new neighborhood
Being called on in class	Being hot, cold, hungry, or tired
Writing a difficult report	Being late for class
Learning a new skill	Being too busy
First day of class	Receiving a report card

Answer these questions.

1. Which of these sources of stress have you actually experienced lately?

   _____

   _____

2. Which of these sources of stress have you never experienced?

   _____

   _____

3. Explain why you find each of the first three sources on your list to be stressful.

   _____

   _____

4. What can you do to lessen stress from your first three sources?

   _____

   _____

Name _____ Date _____

On Target for Healthy Habits

It is important to have healthy habits. It will give you a healthy lifestyle. Take this self-test to see if you follow good health habits. Answer each question *yes* or *no*. Total your scores in each area: physical fitness, nutrition, and health and safety habits. Color in the targets to show how close you are to the bull's-eye.

Exercise

	Yes	No
1		
2		
3		
4		
5		

1. I exercise every day.
2. I do exercises for muscle tone, such as stretching or weight training, at least three times a week.
3. I increase my heartbeat by doing exercises (such as running, swimming, jumping rope, bicycling, playing ball, or dancing) for at least 20 minutes, three times a week.
4. When practical, I walk rather than ride in cars.
5. I enjoy sports or other physical activities during my leisure time.

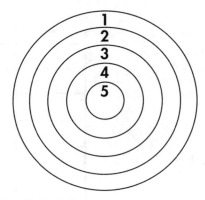

Go on to the next page.

Name _____ Date _____

On Target for Healthy Habits, p. 2

Nutrition

	Yes	No
1. I am at my ideal weight.		
2. I eat three good meals a day.		
3. I seldom eat fatty foods, such as fried foods, hot dogs, and potato chips.		
4. I avoid eating sweets and drinking soft drinks.		
5. I eat fresh fruits and vegetables every day.		

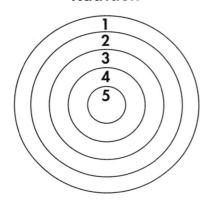
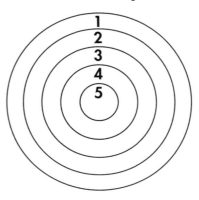

Health and Safety Habits

	Yes	No
1. I do not smoke or take tobacco in any form.		
2. I floss and brush my teeth every day.		
3. I do not drink alcohol or use illegal drugs.		
4. I wear a seat belt when riding in a car.		
5. I am careful when using possible dangerous products, such as prescribed drugs or electrical or motor-driven devices.		

Name _____ Date _____

Good Hygiene

An important part of staying healthy is keeping your teeth and your body clean. It is important to keep the germs and bacteria that come in contact with your body from spreading. This is why you wash your body and hair. It is very important that you wash your hands whenever you handle garbage or raw meats (such as when you make a hamburger) and each time you use the bathroom. Germs can be spread easily from person to person, too. If you cough or sneeze without covering your mouth, germs fly out into the air and onto other people. If you cover your mouth, the germs are contained, and it is less likely that you will spread your germs to someone else. It is not a good idea to share straws, cups, combs, or hats, either. Germs can be passed from one person to the next this way, as well.

Brushing your teeth keeps bacteria from living in your mouth. They eat small particles of food that are in your mouth. Acids can form that eat into the enamel that protects your teeth. When acid eats into your teeth, decay begins, and you will get cavities. So it is important that you brush your teeth after every meal. Just brushing is not enough, however. If you do not floss, you cannot get all the small food particles from between your teeth, and bacteria will grow anyway. If you get gum disease, the gums will not be able to hold your teeth in place, and they will fall out. Visits to the dentist can help keep your teeth strong and healthy, too. The dentist can see problems with x-rays that can be missed otherwise. Then, the dentist can take preventive measures to help your teeth. Brushing, flossing, and visiting the dentist regularly will keep your teeth and gums healthy for many years to come.

Good hygiene helps to keep you healthy and feeling good. It makes you look and smell better, too! Do you have good hygiene habits?

Answer the following questions to see if you need to improve your hygiene.

	Yes	No
1. Do you take a bath or shower every day?		
2. Do you brush your teeth after each meal?		
3. Do you brush your teeth at least twice a day?		
4. Do you floss your teeth every day?		
5. Do you visit the dentist regularly?		
6. Do you cover your mouth when you sneeze or cough?		
7. Do you wash your hands after using the bathroom?		
8. Do you wash your hands after handling garbage or raw meats?		
9. Do you share combs, hats, or other items that go in the hair?		
10. Do you share cups, straws, or other eating utensils?		

If you answered yes to 1–8, and no to 9 and 10, good for you! You have good hygiene habits already! If not, try to improve your habits, and take the test again in two weeks!

Health
Grade Six
Answer Key

p. 3   1. B 2. C 3. D 4. C 5. A 6. B 7. A
p. 4   8. B 9. A 10. D 11. C 12. C 13. D 14. D
p. 5   Answers will vary.
p. 6   Answers will vary.
p. 7   Check students' work.
p. 8   Answers will vary.
p. 10   1. A vegetarian is a person who eats no meat. Some do not drink milk or eat animal products either. 2. A vegetarian diet is usually much lower in fat. High-fat diets are a health risk. 3. bread or cereal: 6-11 servings; vegetables: 3-5 servings; fruit: 2-4 servings
4. Answers will vary.
p. 12   A. 1. d 2. f 3. a 4. b 5. c 6. e B. Possible answers: carbohydrates—bread, spaghetti, candy; protein—meat, fish, eggs, cheese; vitamins—vegetables, milk, fruits, cereals; minerals—all foods; water—water, juice
C. The person skating would use more calories because she is doing more activity.
p. 13   1. 120 without milk; 190 with milk 2. carbohydrates 3. sugar 4. Recommended Daily Allowance 5. 25%
6. calcium, iron, phosphorus, magnesium, zinc, copper
7. wheat bran, raisins, sugar, salt, malt flavoring (Ingredients are listed in order of amounts.)
8. Increase protein by adding whole milk
p. 14   The eggshell in the water was hard. It did not change. The eggshell in the vinegar was soft and easily bent.
p. 16   Acid was working on Kelly's teeth for 140 minutes and on Rita's for 20 minutes.
p. 17   Check students' labels.
p. 18   A. 1. right atrium 2. right ventricle 3. left atrium 4. left ventricle Blood flow: Blood filled with waste from the body enters the right atrium, then the right ventricle. Blood leaves through a large artery to go to the lungs to remove the carbon dioxide and get oxygen. The blood returns to the heart through the left atrium and flows into the left ventricle. From there, the oxygen-filled blood moves throughout the body. B. 5. arteries 6. veins
7. oxygen 8. villi; capillaries; cell membranes; diffusion
p. 19   a. 2 b. 4 c. 5* d. 1* e. 3
p. 20   Answers will vary.
p. 21   1. T 2. F 3. T 4. T 5. F 6. F 7. F 8. T 9. F 10. F
p. 22   1. tall, round seeds, yellow seeds, flowers along entire stem 2. tall, yellow seeds, round seeds, flowers along entire stem 3. tall, round seeds, yellow seeds, flowers along entire stem
p. 23   Combinations: Bb, Bb, Bb, Bb 1. 100% 2. 0%
p. 24   1. 2 2. 4 3. 1 4. 4 C. no D. no
p. 26   1. The eye color would be the same as the father's parents. 2. The eye color would be the same as the mother's parents.
p. 27   Answers will vary based on the results of the survey.
p. 28   2. Answers will vary. 3. Chances for both kinds of earlobes are near 1/2.
p. 30   1. In most cases, the graphs will show the shape of a bell. The height of the greatest number of students falls in the middle, with the least number on the outer edges.
2. The girls are probably taller. At this stage in their lives, girls often grow faster than boys.
p. 31   Answers will vary.
p. 32   1. heredity 2. environment 3. heredity 4. heredity
5. both 6. heredity 7. both 8. both 9. both 10. both
p. 34   1. Price will vary, but should be realistic.
2. cost per pack x 365 3. cost per year x 40 years
4. 20 x 365 = 7,300 5. 7,300 x 40 = 292,000
6. 292,000 x 5.5 = 1,606,000 7. 1,606,000/60 = 26,767 hours; 26,767/24 = 1,115 days; 1,115/365 = 3 years
8. Increased medical expenses (cancer, heart disease, circulatory disease, emphysema, bronchitis), increased cost for life insurance, ruined clothing and furniture due to burn holes
p. 35   1. There is a dark brown spot on the cloth where the smoke was blown. 2. The same dark material is coating the lungs. 3. Like the cloth, the lungs are porous, but it takes longer for the lungs to become as dark as the cloth. 4. Check students' paragraph.
p. 36   Answers will vary based on data.
p. 38   Across: 2. fetal alcohol 3. absorb 4. BAC 7. hangover Down: 1. alcoholism 5. cirrhosis 6. drunk
p. 39   1. Stomach Medicine 2. for indigestion, for nausea, etc.
3. Caution: If taken... 4. Warning: Keep out...
5. Directions: Adults-2 tablespoons, etc.
p. 42   Sample answers given. Down: 1. A liquid drug that is widely abused. 2. Drugs that are depressants, also known as downers. 4. This slang term means marijuana.
5. This drug is found in coffee and speeds up the nervous system. Across: 3. This drug is a hallucinogen.
6. This term is slang for amphetamines. 7. These drugs cause users to see their environment differently than it really is. 8. This drug speeds up the functions of the body. 9. This drug is found in tobacco and tobacco smoke.